FIRST STEPS TO
LEADERSHIP

*Start Leading with the 17
Indisputable Laws of Greatness*

2.0

Wale Adekanla

*International Speaker and
Best Selling Author*

FIRST STEPS TO LEADERSHIP

Start Leading with the 17 Indisputable Laws of Greatness

2.0

WALE ADEKANLA

THE 17 INDISPUTABLE LAWS OF GREATNESS HOW TO START FROM WHERE YOU ARE

Copyright © 2015 WALE ADEKANLA ISBN 978-1506122205

www.waleadekanla.org

Sidney Sanni Publishing. London.

Editor: Uche N. Adi.

ACKNOWLEDGMENTS

To Sam Adeyemi, Smith Bam, Sidney Sanni and all who inspired, encouraged and helped me in writing this book.

To my family members, mentors, and associates around the world who have influenced me and contributed immensely to my knowledge in creating these indisputable principles of greatness.

To all my Leadership Ambassadors:

- You share my vision

- You communicate my standards

- You help Africans to discover, develop and demonstrate their true potentials.

- You transform followers into leaders by equipping others with my leadership beliefs.

Thank you all.

FOREWORDS

Setting a goal and achieving it, is the simplest form of success and everybody can actually be successful. I am all for success but Wale takes it one giant step ahead.

Achieving greatness isn't the norm or the common language on the lips of mortals. Saddled with the quest for survival, the thirst for greatness eludes most creatures of thought but not you. You are holding this book because the book had been looking for you and was written with your rainbow quest. Something in you knows that greatness isn't genetic and it's not injectable. Thus you sought this path and waiting was this book to guide you to the greatness you were born for. Greatness begins when you discover your core, your center and then turn it to a magic zone for the benefit of mankind. So that when you die, you will live on in the minds of those alive. Greatness is the product of following a path, aligning with a series of processes and undoubtedly abiding by natural laws. Playing second class isn't your destiny and you can become great as well by simply learning and applying what made the great great.

Wale has distilled those laws as he sketches each line like an artist and paints them on your heart. Let the earth hear the music from that heart before your stay on earth expires. Don't die like someone that never lived. Why die without ever living? Why close your eye when you truly never saw? Why stop breathing when you merely existed and never lived. Allow this book to remove your name from the list of peasants and let it set you with royalty because that is where you were born to be. Live your destiny, BE GREAT.

—Smith Bam Author and

Motivational Speaker,

Nigeria.

Wale Adekanla has masterfully laid out some key laws and fundamental principles that would guide any person aspiring to achieve true greatness. I highly recommend this book to all current and future leaders that desire to be great.

—Ajukwura Wokomah

Health Safety & Environment

professional, United Kingdom.

True greatness is not in what an individual can do for himself but in what that individual does that impacts the lives of others positively. In the history of our country at this time we truly need great men and women who have compassion and commitment and can sacrifice for the greater good. Greatness is not money, there are too many tyrants who have loads of money but we dare not say they are great. Greatness is not popularity; people who pursue popularity for its sake soon fade into oblivion. Greatness is not lording over others; our continent is full of rulers who rule even though it is against the will of the people. Greatness is impacted. What impact are we making? Not only when we are alive but also when we are no more on the scene. Wale Adekanla has put together a book that will set us thinking and inspire us to take action for the greater good. The world needs more great men and women, this book will show you how. Enjoy! And go and do great things.

—Sidney Sanni

Speaker, Author, Entrepreneur,

United Kingdom.

TABLE OF CONTENTS

INTRODUCTION

You are now holding a book with irreplaceable information that can transform your life. It took me five years of intensive study to acquire specialized knowledge from exceptional and common sources. This book contains a decade of practical experience. Many people have missed greatness because of the wrong definitions that they have learned. This book has redefined greatness and will help you to become what you have been created to be.

The potential for greatness is a common power that lies within every human being in the world. I have discovered that everybody wants to be great, but only a few people know and understand what greatness is all about. Regrettably, there are many philosophies that are constructing people's minds against their true potentials. These beliefs have made some of us tell lies to ourselves that we don't desire greatness. Some of us have been made to believe that the desire for greatness is not normal because of our environments. Due to several delusions about greatness, this down-to-earth material has been written to reveal the divine disposition, and to help you discover, develop, and demonstrate your true potential.

Naturally, anyone who does not wish to be great is not normal. Why? God is not against the desire for greatness. He did not create anyone to be a failure. Every legitimate great person is good for God's glory. Therefore, anyone that does not desire greatness is abnormal because the desire for greatness is divine. Every seed that does not look like the parent fruit lacks originality. You are a true child of God and you need to know who you are. You were sent from the 'country' named 'Heaven' to this colony called 'Earth' with an original intention of re-establishing God's kingdom on Earth. He created both Heaven and Earth and put us here on Earth to manage and dominate. He wants us to function like Heaven and that made Him give us the authority to dominate over everything except human beings. This authority qualifies every human being to be an extension of God and achieve greatness.

Our Father in Heaven is great, and He made everyone in His image and according to His likeness, then He gave everyone the original power to have dominion over all things on Earth. The truth is that we all have the intrinsic-dominant power through God, the Father. Therefore, if you don't think, feel, see, and act like God, you are naturally abnormal because your design and being are like God. This book will help you to change your mental conditioning so that you can begin to behave like the Father.

Well, I am not surprised because it all started from the days of Jesus Christ. There was an argument amongst Jesus' disciples about greatness. They wanted to know the greatest among them. After the dispute, Jesus sat them down and said: "If anyone desires to be the first, he shall be the last of all and servant of all," (Mark 9:35). In addition, Jesus declared a profound fact about greatness in another way, while speaking to His twelve trainees: "You know that the rulers of the Gentiles lord it over them, and those who are great exercise authority over them. Yet it shall not be so among you; but whoever desires to become great among you, let him be your servant. And whoever desires to be first among you, let him be your slave," (Matthew 20: 25-27). Jesus used this leadership class to instruct everyone not to rule, oppress, or dominate people. The two words, "anyone" and "whoever", exemplifies that greatness is open for everyone.

Jesus made it clear that your leadership position in life is available. Note that the idea of greatness was neither condemned nor confronted by Him. Rather, He taught them the pathway to greatness. This is a journey that begins with the understanding that there is space for everyone to occupy. Your first task is to find what you have and serve it to meet the needs around you. The process of achieving greatness includes the ability to discover, develop, and serve your gift to humankind. When you solve problems at a higher dimension, it will shoot you to greatness.

Your uniqueness is in the gift that separates you from virtually eight billion people in the world. This book will help you to recognize, develop, and serve your gift. You don't become great by dominating people. You become great by dominating your area of gifting. This is the point: you need to serve your gift to people if you want to be great. People are not necessarily fascinated with you, but by the fruits from your gift. My conviction is that your gift will introduce you to the world if you use it (Proverbs 18:16-17). The 17 laws will help you to understand this better.

I have met people around the world who pretend as if they are not fretful about greatness, but deep inside them, they wish to become great. Many people are frustrated in their 20's, 30's or 70's because of what they don't know. Terribly, you cannot appropriate what you don't know and understand. Sometimes, they are confused and shed tears inwardly because they don't know how to start or continue. The fact about life is that there are losers; there are winners, and there are people who are yet to discover how to win. You are not meant to die at 18 and be buried at the age of 98. You are here to achieve a divine purpose. This book will show you the key secrets on how to make a forest out of the seed inside you.

Until 2006, I thought I was nobody because I was born into an average family and grew up in a town named Ilesha in Osun state, South West of Nigeria. Everything appeared not to be working. I had numerous reasons why things wouldn't work even without any attempt. I was so trapped in the average culture and later became a strategic inventor of reasons. I never saw any possibility ahead of me. I was completely blind. The world kept showing me day after day as if it was against my existence. I thought that was how God wanted me to be. I incontestably enjoyed the chicken's life. Oh, the chicken's life was not sweet at all. I was living what life only presented to me until I discovered that there are only two fallouts in life: excuses or results.

These laws were revealed through the hunger for spiritual food and understanding of God's mind concerning the creation of man (2 Timothy 3:16-17). I was in a chaotic state, which drove me to look for these secrets. Just like many people are experiencing right now, I had existed for years without the understanding that I am a sanctified being (Jeremiah 1:5). I was living like an alien. I did not know that I have the right to claim the true citizenship of my divine country (Heaven) and that I am an Ambassador on Earth. In fact, it was hard for me to believe that I was born to soar like the eagle. Due to my experience, this book also explores how to bring an eagle out of you.

You need to first know that you are an eagle, not a chicken. You were born to soar and take charge of Earth. You have all it takes to become great within you. Nobody can stop you except you. Nobody can make you succeed or fail without your permission. If you are presently in a condition like a caged eagle and you feel like being in prison because of your background or race or status, your release is guaranteed as you discover these life-changing laws. The potential of an imprisoned eagle is limited until it is released. You need to know how to break out of the prison of limited philosophies and spread your wings.

Let me tell you something now; you need to taste flight. Leonardo Da Vinci said: "When once you have tasted flight, you will forever walk the Earth with your eyes turned skyward, for there you have been, and there you will always long to return." This book will show you how to taste flight. You have the ability to soar like an eagle. Everything, including those things that seem impossible, is possible. The antidotes are inside this book.

It is vital to know that this world is governed by unbreakable principles. These principles have imperishable controlling power. But as forceful as they are, they don't enforce themselves on people or situations. They rather respond to recognitions. That means great things happen when they are only known, understood, and applied. It is time for you to know.

God's intention from the beginning was to make greatness out of everyone (man or woman). He created the fish with its inherent power and then put it in water. He created the bird to fly in the sky. He puts the supernatural power inside a seed for it to become a tree when it is well planted and nurtured. God created everything and positioned them in their natural habitats. If a fish cannot swim, it would place a dishonor to God's reputation; if a bird cannot fly (except those which were designed not to fly), it would put God's name in trouble. Therefore, if you were created in His image without the power to act like God, it brings Him into disrepute. God designed you for success and He will always protect his integrity. So, it is your birthright to be great. However, if you don't search to know His mind for your creation and you don't work with His laid down principles, then failure is predictable.

Evidently, I did not create these laws. I also met them here. God orchestrated the laws so that life can be full of milk and honey for those who know and obey them. For me, I discovered and decided to obey them because of the understanding that every great success has a condition that must be fulfilled. The cheap things are not great, and great things are not cheap. Greatness requests for a price that must be paid.

The Bible, also known as the 'life manual', contains laws and promises. The book is stronger than a legal piece. In it, we have conditional words like "but", "if", "then", and a lot more. If you want to understand how to obey God's conditional words better, study the book of Joshua. Joshua's success with respect to the Promised Land was conditional. God said to him, "This book of the Law shall not depart from your mouth, but you shall meditate in it day and night, that you may observe to do according to all that is written in it. *For then* you will make your way prosperous, *and then* you will have good success," (Joshua 1:8). This simply means that promises are given when commandments are obeyed. There is nothing anybody can do against the laws. Jesus Christ obeyed divine laws and abided in God's love (John 15: 10). We cannot break divine laws without the experience of undesirable consequences.

The laws were received through the revelations from God. Matthew Ashimolowo, the Senior Pastor of KICC ministry said, "Only 10 percent of people are shaking the world." This is shocking! Are you also surprised? These people are great minds in different fields all over the globe. They are who they are because they have discovered and applied these divine principles. That means 90 percent of people are dying without achieving more than 10 percent of their true potential. Also, John C. Maxwell explained that out of the key 300 people mentioned in the Bible; only 20 percent were able to fulfill their purposes. These people understood and conformed to these divine principles.

This book has been written to help individuals, organizations, and nations to discover, develop and demonstrate their latent potentials for greatness. It will expose everyone to the river of greatness in the Heavenly bank account. God has deposited all you need in the bank account of Heaven. As His image, you only need to make withdrawals as much as you want. But the password to access the account must be known and used. This book will communicate the password; how to access the account; how to write divine cheque; and how to make reliable withdrawals. The more you discover and act, the more rapidly your withdrawal capacity would be.

It is imperative for you to know that the irrefutable law of gravity cannot be broken. If a man prays and fasts for 40 days as Jesus did and then jumps from a 10-story building anywhere in the world, believe me, he would hit the ground, bounce like a ball and may likely die immediately. Laws are no respecter of your person, background, race, religion, or past. If you know and obey them, they will work for you, but if otherwise, you are doomed.

Laws are not meant to be experimented with because the result of disobedience is unendurable. The manufacturer of an iron device includes in the manual that you should not plug it in water. If you do, I am quite sure that the story won't be sweet. The laws respond to obedience. You don't test laws, you obey them. God designed them for us only to be discovered and obeyed to achieve His expected ends. The laws make success predictable and confirm the possibility because they are well-founded.

This book is also packed with the incontestable and traceable secrets of God from the beginning of all creation. The beginning of everything is as important as the end. If what you know and do cannot be drawn from the beginning, they will never lead you to greatness. But if they are connected to the mind of God as designed from the beginning, then greatness will become inevitable in your life.

Every beginning that God has not authored will lead you to a lifetime of failure. You have been designed to be great but cannot automatically become so until you choose to be. You were born great, but growth into greatness is a must. Therefore, you have to discover who you are; what to know; how to do things; and when to get things done.

You cannot buy the most important things in life. You can only buy things that are less important to God. But when you have the things that money cannot buy, then you have the right to possess the things that money can buy. For example, you can buy an expensive and customized bed, but cannot buy a good sleep. You can buy a mansion, but cannot buy a home. You can buy a companion with your wealth, but cannot buy a great friend. The understanding of these laws will give you wisdom and brainpower that money cannot buy and brand you to become what you have been designed to be. You don't have to be great before you start, but you have to start before you can achieve greatness. It is time to start.

The contents of this book are very simple to read and understand, but never try to finish it in just one sitting. Focus on a law each day. Your determination to commit yourself fully to the principles taught in this book would determine the overall worth of this timely investment you are making. With your notepad beside you, I am convinced the journey can now begin.

Welcome to the uncovered secrets of greatness!

CHAPTER 1:

THE LAW OF GREEN THINKING

*"You naturally grow living thoughts when you open
your mind to worthwhile things."*

The mind is the prearranged original source of every great success. Real success is inside out. Every great visible thing that exists today started from the mind of someone. Everything that exists in the natural started from the supernatural. Even God thought of creation before He created. He said, "Let us make man in our own image." This discovery has revealed to me that the hardest and the most expensive work in the world is thinking. Thinkers are not cheap people. They are not everywhere. They expand their minds and get excited with passion in the presence of problems. They sit down and think. Thinkers make an impact and become real leaders.

Green Thinking is not lifeless

This kind of thinking has life and produces unusual results. Worthwhile thoughts are full of life. As a person, the hardest task that I have ever done in my life is the rebuilding of the human mind. If you can influence someone's negative way of thinking, then you have successfully changed that person's life. The summary of the devil's struggle is to influence the human mind because of his knowledge. He knows that the heart is the most critical faculty in the body. The first project Jesus carried out when He began His ministry at the age of 30 was on the human mind. His first public declaration was, "repent!" Repent as a Greek word means "change your mind". He was precise about the mind. He realized that the greatest challenge was the wrong conditioning of the minds and that made Him declare His mission statement (Matthew 4:17).

The fundamental difference between a rich and a poor person is the way they think. Let me explain that. A rich person thinks rich while a poor person thinks poor. A rich person thinks he is rich even when he has nothing of physical value. My close relationship with people has made me know that everyone wants to live a better life but not everyone thinks better. A poor person wants a better life, but he knows that he is poor and accepts it and that is the core problem. The knowing and accepting are simply called 'self-image'. This 'self-image' is basically the cause of poverty because you can't get anything different from what you think. You are ultimately what you think. Our lives can never be better than how we think because it is an unbreakable law.

How we think controls our decisions and choices in every situation. Sometimes we get confused when people don't do the things that we want them to do. The best way to understand how someone thinks is to focus on the choices he or she makes. We are what we think. Why? Because we don't get what we want in life, we get who we are. Decisions are made through our thoughts and every decision we make confirms or compromises our values.

Thoughts Become Things

Another key lesson is that our thinking helps us to attract who we are. Think right and you will attract the right people, think wrong and you will attract the wrong people. You have to be sensitive to what goes on in your mind and who you are becoming every day. If you want the best for your life, you have to become the best yourself. However, if you want to become the best, it must start from the mind.

My interest in nature and the animal kingdom made me study and discover the leadership secret of the lion. I got stuck and went into deep thinking when I was studying this animal. In the animal kingdom, the lion is not the largest; not the smartest, not the tallest, not the longest, but the lion is the king of the savannah. This is absolutely mystical. I discovered that the secret behind the lion's leadership is attitude. The lion knows its identity in its habitat. Lions are bold, powerful and compelling animals in all situations. If a lion and an elephant meet, the lion's natural thinking would be 'lunch'. The thinking faculty of a lion controls its behavior in its habitat. The lion is the king because of its thinking. The lion's attitude is different. Truly, attitude is everything. How we think determines how we behave.

The best way to have control over our lives is to consciously manage what we think because thoughts become things. We cause every effect of existence through our thoughts. It starts from your self-image, then self-esteem, then self-worth, and then self-concept. Your self-concept is your philosophy and that makes you act unconsciously. The effect is your self-concept while the cause is your self-image.

How you think reflects the way you feel, talk, and act.

Let me share a simple poem I wrote:

THE MIND!

There is only one house of battle

A place where everything is caused

House of failure and success creations

Equipment that processes the invisible

A bosom friend of the universe

There is only one house of battle

That place where every old and young fight
every second

The house of mystery that holds tomorrow's
blueprint

A factory where tomorrow is manufactured from
today

An engine that dictates the future

There is only one house of battle

A place to make a personal choice - success
or failure

The converter of raw materials: Thoughts
into products (Realities).

The occupant that connects the open gates

Receiving from individual Eyes and Ears to
program

There is only one house of battle

The field of battle for the two ever- strong giants

Everybody wonders how it feeds - Positive or Negative

Starving giant loses the battle and shifts for the winner to reign

The mighty internal mansion for success

Thoughts are powerful tools for all creations in the world. Your thoughts can block or open your life to anything. Get it once and for all; your life tomorrow can never be better than the quality of your thoughts today. Thoughts are seeds in your mind that turn into realities. What you think is what you get (Proverbs 23:7). That is it. If you think something is impossible even before you start, that is exactly how it would be.

Poor background, poverty and slavery mentality can be changed to success and significance through our thoughts. The first thing is to change the way you think. Don't put yourself down in your thoughts. You can only be a billionaire tomorrow if you are one in your mind when you have nothing. Your situation now is not the determinant; it is your attitude to the situation. Attitude has the potential to build or terminate your life. The problem is not the problem; the problem is the way you think about the problem. You can only change the name of your family tomorrow if you change it first in your heart today. Possibility starts from the mind. It is a battle you must win.

Green Thinking is Possibility Thinking

There is a connection of the mind with the brain. Neurologists understand this very well. The brain does not only understand any information sent from the mind, but it also obeys it. Information from the mind is a command for the brain. So, you must protect your mind gates. Disallow jargon to enter through them. Watch what you see, hear and say. Then, note that no one can think differently and not achieve different results. I love to say, "If you can think like one in them and not one of them, you will achieve a result like no one of them." Thinking is hard work. Popular thinking is not possibility thinking. Green thinking is possibility thinking.

The reason many people don't think and make big things happen is that the majority believes that thinking is not working. Through my experience, I have come to realize that thinking is hard work. I encourage you to think and think big that all things are possible for you, even from the slum. But if things are good for you, there is another place that is better than good. Be transformed on the inside and you will achieve greatness.

CHAPTER 2:

THE LAW OF DISCOVERY

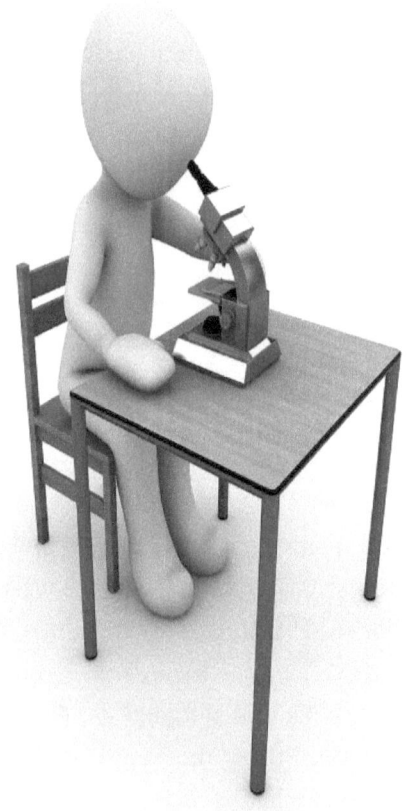

"You don't create the purpose for your creation, you discover it."

The greatest question that gives meaning to everything on Earth is, "Why?" Have you ever asked yourself this question, "Why am I here?" One important truth you need to understand is that you are not here by mistake. Jeremiah 1:5 describes God's intention before you were created just the way you are. There was a plan in heaven before you were formed. God had something in mind before He sent you to the planet Earth as a human being but not as a dog. God had a motive for creating you the way you are. A meeting was specially held just because you were coming to this world. He knew you before you were conceived. He has not sent you for yourself. A journey began the day you were born. Now, have you discovered His intention for your creation?

Your Advantage

You gain no advantage for going after what your life's purpose does not demand of you. It is extremely important to focus on what you were born to do. We all know that age increases every day and that is why people don't ask us "How young are you?" They rather ask, "How old are you?" The day God sent you to this world; you automatically received a return ticket. That means you are on a journey and you must return. You need to answer these questions: Why are you on this journey? Why are you here at this time?

You were not created to only work for 40 hours per week and retire after 40 years. That was invented by man, not God. There is nothing like retirement in God's program for your life. The word 'retirement' doesn't exist in the Bible. You are here for a purpose. Your purpose is your destiny. It is tragic that some people in their 80's don't know why they are here. They are succeeding in the jobs that the creator didn't send them to do. They are living miserable lives and still asking unanswered questions in their hearts. Martin Luther King, Jr. said, "If a man hasn't discovered something that he will die for, he isn't fit to live." Now, my question is, "Are you fit to live?" You are loaded with an assignment that is in the form of potential, waiting for discovery. The gift in you is seeking expression. It is time to discover who you are and live a full life.

Every Good Seed has a Forest in it

An ultimate fact in life is that everybody has the seed of greatness on the inside. No one came to this world empty. No matter your background, temperament, race or status, you have this special gift that is called 'The Seed of Greatness'. The natural phenomenon about this seed is that it has the potential to produce a forest. The word 'potential' indicates that the seed has the ability to become a forest, but that doesn't mean that the seed would be. It has to undergo certain changes.

The seed will never grow if not discovered, planted and nurtured. Let's assume someone gives a good seed to a farmer, and he plants it at the right temperature and humidity; beyond doubt, it would germinate, grow and become a tree under the right conditions. Apparently, on that tree, there would be many fruits with the potentials to produce more trees if planted. If the right planting continues, the one seed would give birth to a forest. The farmer's seed represents the seed of greatness within you and the capacity of its potential. This seed is found in your natural habitat. It is a gift that distinguishes you from other people in the world. However, it has no capacity of its own to grow until you make it happen. The gift must consciously be positioned in the right environment; must be given the right nutrients; and must develop through the process.

In my early twenties, my father trained me on how to plant and nurture good seeds. I discovered a lot from his words. He taught me how farmers do search for good seeds to plant, calculate the yearly seasons, undergo regular weeding, and plan to have a good harvest. The same process applies to our lives. What is the seed of greatness in you? The greatest mistake people make is that they think they have nothing inside them, so they live like nobodies. You have something special in you like no one in the world. You are unique. It is not your responsibility to create the seed, it's already inside you and waiting for you to discover, develop and demonstrate. The late Myles Munroe said, "There is something for you to start that is already destined for you to finish if you can just discover."

Meet Yourself

Leo Buscaglia, who was an American author and motivational speaker, and a professor at the University of Southern California said, "The easiest thing to be in the world is you. The most difficult thing to be is what other people want you to be. Don't let them put you in that position." Many of us have met several people in our lives, but yet to meet ourselves. You can only live the real you when you meet yourself. You can meet yourself when you discover who you are. I want to put you through a personal practical program (PPP) of self-discovery by using the 1969 Johari's Four Windows, which was named after the first names of its inventors, Joseph Luft and Harry Ingham.

JOHARI'S WINDOWS

	KNOWN TO SELF	UNKNOWN TO SELF
KNOWN TO OTHERS	OPEN	BLIND
UNKNOWN TO OTHERS	HIDDEN	DARK

These four windows are quite simple to apply and experience in personal awareness. They will help you discover some secrets about yourself. Let's look at them one after the other:

Open Window: This represents the things you can do and others know that you can do it. Everyone including you knows that you have the natural skill or gift. Let's say you can easily organize events or calculate things, which are known by you and others around, which undoubtedly is your open window.

Blind Window: This stands for the things people around you know that you can do, but they are still unknown to you. These are the things you own but have not discovered. People know that you are skilled, but unfortunately, you do it innocently. This happens simply because people can see you more than how you can see yourself in the blind window.

Hidden Window: This is the zone that explains what you know about yourself that others don't know. The potential that only you are aware of will never be exposed to people around until you unveil it. Many people are trapped in this zone because of fear of failure or their mediocre culture.

Dark Window: This zone defines the human complete ignorance. It is what I have called 'Inside out Ignorance'. Your gift is neither known to you nor to people around you. No one knows the power in you and that includes yourself. Here, you need to seek the face of your Manufacturer before you can experience a glimpse of who you are. God has the answer to every unknown in your life because every manufacturer knows the all-inclusive functions of his products.

The definition of a complicated life is found in the ignorance of it. It is disheartening that many people are ignorant of who they are, just because they lack the original information. They carry the princes' potentials and remain in strange kingdoms without identifying their kingdoms. Unfortunately, a prince who is a potential king will never become a king if not positioned in his own kingdom. The only way to live your making in life is to identify the purpose of being created and stay there. Pray and think about why you are here. Are you presently living or existing? You need to answer that as a person. Don't deceive yourself because I know that you can do great things. Discovering your reason for creation makes you live your making, not making a living. It is not the issue of where you are, it is about whom you are and what you are capable of becoming.

Here is my poem about personal discovery.

Why Are You Here?

The greatest question in life is not "what"? The greatest question in life is "why"?

There is a space for "Why" in every "What"

The purpose of a thing is in the heart of the maker.

Nobody knows Toyota like the Company.

Nobody knows Mercedes Benz like the Company.

Nobody knows Ford like the Company.

Nobody knows you like the Manufacturer (God).

People only experiment when the purpose is not known. Everyone apart from your Maker guesses your purpose People abuse you if you don't act your purpose.

Leadership starts when the purpose is known.

Everything on earth has a purpose.

God created you for a specific assignment. God had an intention before He sent you to this world.

Why are you here on earth planet?

When will you start to Deploy?

You cannot start, continue and finish the work you know nothing about. Regrettably, many people are very busy doing what they have not been wired to do. Sometimes, we call them lucrative jobs. Nothing is wrong with working for someone for a period, but something is wrong if you die without living by intention. Your job is what you do and receive a paycheck for, but your work is what you were born to do. Never wait to be retired before you leave if you want to live your dreams. The earning from your job is meant to be used for the development of your work (purpose). The deployment of your gift is more important to God than your employment.

It is sickening that the universal culture focuses on employment, not deployment. We are not here to focus on jobs; we are here to focus on our works. You are employed when you are on a job, but you are deployed when you are doing your work. Deployment is serving your discovered gift to the world. You can never retire from your gift. Greatness shows up when you find your gift. It is important to transition from self-dependence to self-independence.

You have to discover and finish the work God has sent you to do. When Jesus was done on earth, He prayed a prayer for Himself and said, "I have glorified you on earth. I have finished the work which you sent me to do" (John 17:4). His gift was to give salvation to the world and His destiny was to die on the cross. He actually fulfilled the re-introduction of the Kingdom of God. He knew why He was sent and finished the work. Are you going to finish your work?

It is necessary for you to think about who you are and ask God for help. When the sum of your mental, psychological and spiritual out-sources is combined with your in-source, the bulk is capable to produce leading results in life. Discover your purpose and start living the reason for which you were created.

Heroes and Heroines understood the concept of Purpose

There are many great Heroes and Heroines in the Bible like Moses, Joshua, Joseph, David, Esther,

Daniel, Mary, Paul and so on. To talk about Jesus, He discovered who He was at the temple in Jerusalem. He decided to sit in the midst of the elders and great teachers. There He listened, observed and asked questions. To the surprise of the scholars, His response to questions indicated that His reasoning was beyond His age. He talked like an adult because He knew what many adults did not know.

Something noteworthy happened when His parents found Him after the anxious search for three days. When His mother told Him how they had been bothered about Him, He humbly told them, "Why did you seek me? Did you not know that I must be about my Father's business?" (Luke 2:49). The powerful and unexpected response from a 12-year old child was a shock. He had discovered why He was created and sent to the world. He understood God's intention. This is exactly how you are meant to live your life because your purpose is God's intention for creating you. It is to be discovered by you, not to be created. When you discover your purpose, nobody on Earth would be able to miss-use you. Although, daring your purpose may disconnect you from some people or lead to temporary appalling circumstances, but no matter what, you must do your Father's business. No one lives a purposeful life without good success and fulfillment.

When Jesus first spoke about His God-given goals, His parents did not understand the statement. Do you know why? It was difficult for them to comprehend

His divine status. You were also specially made for an exceptional purpose which your family or people around you may struggle to understand. They are mysteries which only God can reveal to individuals. If some people don't understand or believe in your reason for living, it's not your problem. The greatest problem is if you don't know who you are and believe in yourself. Nobody in the world can stop you if you don't stop yourself. Therefore, a decision to discover who you are is paramount in achieving greatness.

Evidently, you are here for a purpose. If you don't know your purpose, connect with the Maker. All manufacturers understand the purpose of their products. God, the manufacturer of every man and woman understands 'why'. Never try to plan your life before the discovering of your purpose, in order not to struggle on the wrong road (Proverbs 19:21). Find your purpose and then plan to achieve it. It is better to do nothing and know why you are doing it than to be doing something and not know why you are doing it. Discover why you are here and live a reasonable life.

CHAPTER 3:

THE LAW OF SERVANTHOOD

"If you hold service as a key, you will open the door of greatness."

The Best Leadership Class

The surest and the only way to the greatness that Jesus taught His disciples is *service*. He described a process of leadership for them through servant-hood. He taught them to become servants of all.

However, working has led me to discover that many of us have misunderstood Jesus' concept of greatness. The first thing is to understand this idea of leadership and disconnect from the traditional leadership concept. The word 'servant' makes us think more of being lesser than someone. Most times, we think more of slavery instead of serving one's gift to the world. Servitude was not what Jesus taught in that leadership class. He was talking about the uniqueness of your gift, and that if you don't give it for the benefits of others, then you have deprived the world of your irreplaceable contribution. Jesus did not mean to say make yourself less, but be significant by distributing yourself to others.

Who Are You?

I am convinced that everyone was born with the potential to lead and become great, but this is not automatic. The orange seed has the potential to become an orange forest, but that doesn't mean it would be. The seed must first pass through a process of development. We can also say every newly born male has a man, father or husband inside him. In the same manner, the seed of greatness inside every human must be developed and served to others before the achievement of greatness.

Jesus discovered His gift when He was 12 years old. He challenged His parents with a question, "Why did you seek me? Did you not know that I must be about My Father's business?" (Luke 2:49). That was a sense of recognition. His leadership was born at that age. That statement showed that He discovered something important. He discovered His gift. He was able to find His personal purpose that must be served to others. This is very important for you as well. You must discover who you are and distribute yourself to the world. This is what I call real freedom because you will be free from culture and religion and become what you have been created to be.

Greatness is about self-discovery, self-development, and self-distribution. No one in the world can place you on your seat of greatness. You cannot attend a university and study a course and just become great. You can only become great when you discover the power in you, your gift, humble yourself as you develop and serve it to the world.

You can Become Great

Every seed is a tree and the tree's fruit is a gift the world. But the seed must pass through a process before it can become a tree. You need to be ready for greater challenges. The process of greatness requires the courage to move on. The seed needs to first die and then rise to grow before it becomes a tree. Jesus was arrested, and he died for us without committing any sin. He was greatly exalted by God. Mother Theresa quitted her job because she wanted to serve the poor in India. She later sat with presidents of nations, popes and kings. Martin Luther King Jr. stood for others and demanded equality in the United States.

David discovered and developed his gift of using the sling. He used that to bring Goliath down because he resolutely believed in his gift (1 Samuel 17:50). He rejected the outdated leadership concept when he was offered the armor and sword. He served his people with his discovered inherent gift. I want you to know that you are gifted. Your gift is your leadership domain. Your leadership domain is your power. Jesus made it clear that greatness is open to *everyone or anyone* that wants it when He used the word *"whosoever"* (Matthew 20:26). The word 'must' in His statement means servant hood is 'required' or 'compulsory' before greatness. Also, the word servant means 'meeting needs' or 'serving your gift to others'.

Create and Distribute

People ask me, "What is the practical way of serving one's gift to others?" This is simple and straightforward. First, understand that you have to discover your gift and then have a development plan before serving it to others. To serve, create products or services that can meet people's needs from your gift. You are serving people when your products or services are useful for others. Your capacity for service will determine your leadership level. If you serve a small domain, your leadership would only be felt in that area. But if you expand and distribute your products or services to different domains, then your leadership level would increase. It is possible to influence the world with your gift.

Do you know why the Dead Sea remains dead? It doesn't flow out. Nothing is alive in the Dead Sea. People who don't give are as dead as the Dead Sea. The most important thing that we all need to know is that life is not just about receiving. It is about giving. We all need to be a bit like the Sea of Galilee. So, what do you need to do in order to flow and remain relevant wherever you are? Create products and services from your gift and distribute them. Give people what you have. Givers become leaders. Leaders give out their gifts.

Do the Business

It is high time to know who you are because leadership is everyone's business. You need to ask yourself: "What is my purpose?" "What is my gift?" "What can I do?" "What makes me smile?" What makes me cry?" "Am I ready to serve others?" People who are close to you including your parents may likely say, "My dear, you will be broke if you continue with this career. I am not sure that this is the way!" People can help you to clarify your purpose, but you *must* answer the questions for yourself. It is your responsibility to discover why you are here on planet Earth and manifest accordingly.

CHAPTER 4:

THE LAW OF PASSION

"You get the best out of life when you are positioned in your zone of strong thirst."

The most significant way to experience fulfillment in life is to be located in your natural habitat. Let's go through these four questions: What makes you sing? What makes you cry? What makes you dream? What change hits your heart the most where you are? The answers to these questions express what you are passionate about. The house of passion is the heart (Matthew 6:21). It is the only fuel or drive needed by your talent to function at its peak.

Everything created has a natural habitat of performance. It is also called the kingdom or domain. The fish explores and enjoys living in water, the bird flies and lives its making in the sky, and the best comes out of a lion in the savannah.

Think of anything living, there is a natural territory designed for everything. The only place where you can become your best as intended by God is in your kingdom. Identify it and stick to it.

A Call for Fulfillment

Do you complain about your present job or you don't even know how to identify your work? This chapter will help you. It is better to focus on your area of brilliance than your area of competence. Your passion gives birth to brilliance. But, whenever passion and talent meet, greatness is expressed. That is a specific area where you can easily become celebrated. There is something you love to do always that gives you internal joy whether you are paid or not. You enjoy it once you start doing it.

Brilliance is more than just mere competence. You can learn anything and be competent. I studied Chemistry, but I didn't learn the spirited desire to teach others how to discover, develop and demonstrate their latent potentials. I first discovered it and then began the development. Passion is not what you enroll for in a school to acquire. You are naturally wired with it. It is connected to the problems you complain about often. These problems frustrate you anytime you come across them. They always seek your response because you have the natural inbuilt ability. What you want is not available where you are because you have the potential to create it.

What connects you to your heroic future and fulfillment? Passion! It sends you easily to your world of destiny. Steven Jobs, the soul of Apple talked about passion constantly. He said, "Your time is limited, so don't waste it living someone else's life. Don't be trapped by dogma – which is living with the results of other people's thinking. Don't let the noise of others' opinions drown out your own inner voice. Most importantly, you need to have the courage to follow your heart and intuition. They somehow already know what you truly want to become. Everything else is secondary." Also, Jeff Bezos, an American business tycoon, and investor once said, "One of the huge mistakes people make is that they try to force an interest in themselves. You don't choose your passions; your passions choose you." If you cannot see what you want, passion inspires you to create what you want to see.

Passion Opens Doors

Passion makes things easy and sets you aside at work when everyone says it is hard work. It opens your eyes to see the problems that you are destined to solve. What you are passionate about could be sourced from your value, interest, skill or dream. Think about it.

You need to position your life where your heart belongs and be an expert in one thing if you want to perform at the peak. Do everything you can to have relevant information in that area as a lifetime's work. If you are in a job that you don't love presently, I encourage you to re-think and divert because you cannot be the best there.

There is a special internal drive produced to overcome all challenges in your area of passion. Nothing gives you fulfillment like what you love to do. So, tap into it from now on and strive to be the best you are meant to be. Passion expands your ground of possibilities to the greatness you can ever be in the world.

CHAPTER 5:

THE LAW OF INSIGHT

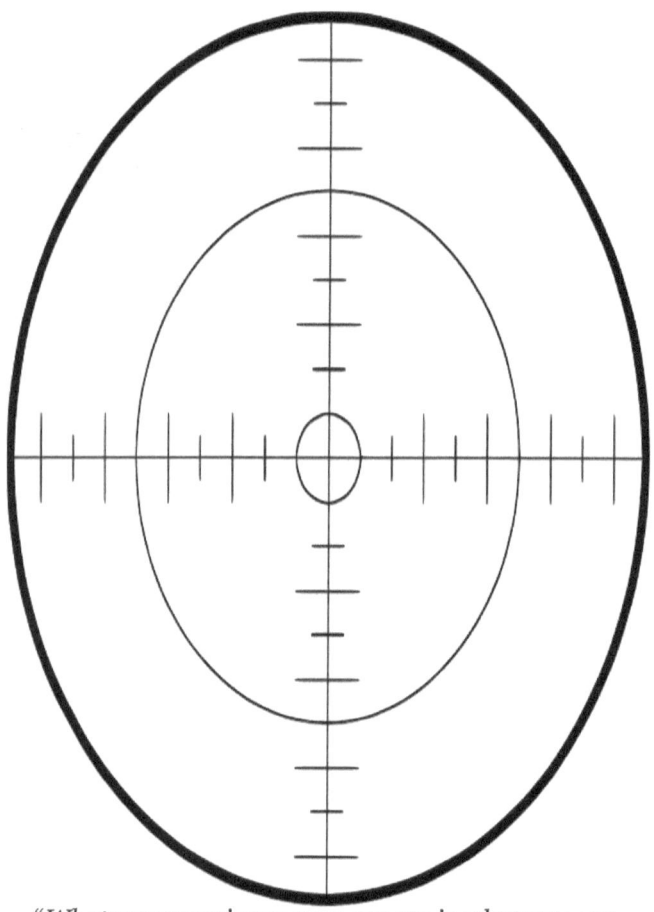

"Whatever your inner eyes can see is what you can get."

The greatest power you have is the ability to see the unseen. Also, the greatest mysterious authority that God has given mankind is not the power of sight, but the power of vision. What you see today in the invisible, in your mind is exactly what you will see tomorrow in the visible. You need to make the short journey in life and be anywhere you desire to be anywhere in the world. The Lord asked Jeremiah the question twice: "What can you see?" (Jeremiah 1:11 and 13). God would not have repeated the question if there was no importance attached to it. James Whistler, an American-born British-based artist who influenced the art world through his artistic theories said, "An artist is not paid for his labor but for his vision." It is time for you to see beyond the present.

What Can You See?

Your purpose creates your mission, and from your mission, a vision that you can see is produced. Vision is about traveling beyond the physical. You need to travel in your imagination into the future. Have you ever visualized your life in the next five or ten years without concentrating on the distracting circumstances around you now? I mean by consciously neglecting that you are poor or average in all things. If you desire to move on, close your eyes and ears because you need to first disconnect from all impossible and negative facts. *The moment you dream, you rise above your limitations and no one can stop you again except you choose to discontinue.* Where will you be in the next 15 years? Write it down. You need to be a dreamer to have a future full of possibilities. When you fail to fulfill your vision, you have failed in your mission. And if you fail in your mission, you have failed in fulfilling your purpose. What have you seen in your mind today that does not exist for you now? The invisible forces are waiting to rush into your life and turn your dreams into reality. God serves every vision with its provision which is implanted in His grace.

Naturally, we believe what we see physically, but we supernaturally become what we can see mentally. The question is "what can you see or perceive"? What you can see about your tomorrow from now is your vision. If you don't have a vision, you are ruined. The significant thing about vision is that it takes you out of all limitations of your physical eyes and gives you the liberty to recognize what your heart can feel. Vision is the bedrock of great inventions. Vision activated the global Microsoft Empire out of Bill Gates. Vision triggered desk-sized computers out of Steven Jobs. Vision is the divine tool that motivates you to make a great impact on your generation. Vision is the unseen tomorrow that you have never lived. Vision is the fundamental foundation towards fulfillment in life. Your vision is your future.

Joseph saw Greatness

At 17 years, Joseph had dreams about greatness. He was the youngest of Jacob's twelve sons. As a boy who doted on by his father, he told his family members about both dreams. In the first dream, he said: "Please hear this dream which I have dreamed: There we were, binding sheaves in the field. Then, behold, my sheaf arose and also stood upright; and indeed your sheaves stood all around and bowed down to my sheaf" (Genesis 37:6-7). Then, he also had the second dream and said: "Look I have dreamed another dream. And this time, the sun, the moon, and the eleven stars bowed down to me" (Genesis 37: 9). He dreamt of all these in his father's house. He saw the future that he would become a Prime Minister in Egypt when he was a young boy. He had nothing, but he saw something.

However, Joseph was hated when he shared his dreams. The first dream made his brothers hate him because they felt threatened by his words. The second dream made his father rebuke him in the presence of his brothers. Jacob said, "What is this dream that you have dreamed of? Shall your mother and I and your brothers indeed come to bow down to the earth before you?" (Genesis 37: 10). In fact, that was the beginning of all the challenges he experienced in his lifetime. In spite of all the challenges he faced in his father's house, in Potiphar's house, and in the prison, the dream came to pass because he kept it, utilized his giftedness and never compromised.

It is necessary to get the losers out of your life if you want to live your dreams. You need to know that everybody won't see what you can see. It is crucial to know that everybody won't join you in your dream. Everybody won't believe your dream. But, never allow people's reactions and negative opinions to become your reality. It's up to you if you want to live your dream. You've got everything you need to live your dream within you. God who gave you the dream will also make ways for you to live it. Never negotiate and don't give up on your dream. Don't give in to any discouragement. Give all you can to your dream every day. Live your dream and experience the real you because you are needed at the top. I encourage you to leave the crowd and move to your greatness.

Working out your Dreams

If you want to help your knowledge of the application of these principles, read the Bible and life-changing books. Reading was recommended by Jesus Himself (Matthew 19:4). Great books have helped millions of people to become great all over the world. They are irrefutable secrets of great people. Great secrets are always packaged and covered. The only way to discover them is to search, open and explore. Nothing empowers you like knowledge. I love books because their contents help me develop continually. They uncover secrets.

Be hungry for knowledge and invest in books because secrets to convert your dream to reality are embedded in them. Books are living things that can fill your heart and transform your life. Your life cannot be better than the knowledge you have. Books empower you to see and act your future clearly from today.

Seeing tomorrow from today helps you to know the directions of your life and practically move you there. You need to see what God has prepared for you in the future with your heart and not merely your physical sight. Believe that it is possible, even if you are passing through hard times. Nature sees those who can see and pulls what they see to them. Without a dream, you have no future. I say to you, have a dream and follow it.

CHAPTER 6:

THE LAW OF PROPHESY

"You command new things into existence each time you speak."

The law of prophecy is the principle of confession. One thing is to believe in your heart that great things will happen and another thing is to say it out. During my study about how God created the world, I focused on chapters of creation; Genesis 1 and 2. I discovered that God did not use physical material to create anything in the world. This caught my attention and made me search further on what He used. I was desperate to know His secret, and I found it. Are you also interested in knowing His secret? Here is what I discovered.

Words have Life in them

In the beginning, we could ascertain three repeated words, "Then God said", in Genesis 1:3-29, to confirm how God created everything. When God wanted to create the world, He had the thought and spoke. He said, "Let there be light and there was light." He declared what He wanted to see, and it happened in the physical realm. This is the exact power God has transferred to us. He created us in His own image so that we could become co-creators. He made us like Him so that we could think, speak and act like Him. So, you have the ability to give life to situations and circumstances. Sometimes, we have negative thoughts but we should not give life to those negative thoughts. When we speak, things may not happen physically like it happened when God created the world, but things are created in the supernatural. This is how God has designed it.

You must learn to control your tongue. Use words that inspire results and not words that will put you down. God saw darkness but called out light and there was light. You give life to every thought you speak out. It is possible to have faith that can move mountains, but if you don't speak positive faith-filled words, unfortunately, nothing will happen. I encourage you to speak out about your dreams.

When I had nothing, I began to declare what I wanted to see. My mother heard me one day when I said, "I am going to the world to fulfill my destiny" with excitement. She could only nod and said, "It is well". I quickly went after her to discuss what I was seeing because I thought she might think I was not myself. In the natural, there was nothing. But in the supernatural, I could see plenty. In fact, it is well.

The Winning Recipe

This is the winning recipe; dare to affirm your dream through your words. One day in 2006, I was in my room and stood in front of a mirror, thinking and speaking alone with high energy and excitement. All my words were the direct opposite of the undesirable circumstances around me. The attitude I demonstrated was more important than the facts that I could see. When I discovered this law, then I made up my mind to practice what God did in the beginning by changing my world with my words. Can you imagine what I did? I named countries I would like to visit without having the money for the first flight. This would be a crazy thing to do for an average mind. I never knew how it would happen, but I kept saying great things everywhere. Today, I have been to many countries, sharing my ideas, because I said it. The words I spoke have come to pass. Now, I speak greater words

Like I did, you have to also stop saying, "I am having a bad day". Say, "I am having a super fantastic day", "I am creating my great future", "I am better than good", "Today is a great day", "I will make it", and lots more. Be assured that God does not create bad days. Every bad day is already influenced by the devil. But the devil and his angels understand the power of words and so they work tirelessly to make you think and speak negatively and attract destruction. As powerful as the devil is, he does not have the authority to create any day. So, never say, "I am sick", "I am poor", "I am nobody", "Things are not working", "Everyone is against me" or "I am not a leader", even if you think it. Those words must never come out of your mouth. Every word you speak creates life or death. "Death and life are in the power of the tongue, and those who love it will eat its fruits" (Proverbs 18:21). The question is "what kind of fruit are you producing"?

Speak like God

Sometimes, I wonder how some people think. They want God to communicate with them without speaking the language that He understands. They have memorized almost all the worldly songs and movies without any verse of the scripture in their brains. They speak the language that God doesn't understand. If you speak Satan's language, he will definitely respond and attend to your requests. The devil wastes no time, he is always looking for whom to devour. You have to think and communicate like God anywhere and everywhere. When the devil tempted Jesus after His fasting for 40 days and nights, Jesus told him to his face, "It is written: Man shall not live bread alone, but by the word that comes from the mouth of God" (Matthew 4:4). Always speak, "It is written", and win the battles against your dreams and achieve greatness.

You are also a co-creator with God, whether you are a believer or an unbeliever. The reason things are bad may likely be that you have sown negative words. Your positive words disconnect you from the devil and activate your faith when nothing is happening. Just speak great words because they will come to pass. Give life to your great future, business, health, certificates and everything you have through your words. Don't say what you see; instead, say what you want to see. Words are seeds. What you say is your choice. Stop complaining. Change obstacles into miracles with your words. Control your mood and intentionally speak positive faith-filled words to change your status quo.

CHAPTER 7:

THE LAW OF STEADY GROWTH

"Daily success accumulates to become an endorsed success."

Greatness is not a day tour; it is a journey. Unfortunately, some people wish to acquire everything they want in a day. I have seen and heard lazy people daydreaming and praying to achieve helicopter success. The helicopter does not have to taxi before it takes off like the airplane. However, an airplane gets to the runway before it takes off and flies high in the sky. Real success is a process. Real success takes off like an airplane. When you grow up, you stay up, but when you jump up, you come down. You need to develop. You need to know and obey the law of steady growth.

The Bamboo Tree

Let me share a great lesson from the Chinese Bamboo tree. This tree has its own uniqueness that makes it different from other trees. When a farmer plants the seed, it doesn't outwardly grow an inch until the fifth year. The farmer has to water and nurture it every day and throughout the seasons of the year. This process of irrigation and absolute care for the seed must be committed to, even when the seed fails to sprout. Sometimes, the farmer would be discouraged because of no results to show for his efforts.

Suddenly, in the fifth year, something strange and mysterious happens. The bamboo tree sprouts and grows 90 feet tall within six weeks. This is interesting. If I may ask you; does the bamboo tree grow for five years or six weeks? Does it decide to grow in the fifth year?

Apparently, bamboo tree seed grows for five years. It expands its root system inwardly and holds the ground firmly waiting for outward development in the fifth year and years after. I encourage you to follow the process because if you arrive all of a sudden without God's involvement, you will depart suddenly. We are all 'human beings' not 'human been'. Being is in continuous tense, which means we are all on journeys. God has the capacity as the omnipotent One to create you as an adult, but He did create everyone as a baby because greatness takes a process. Stop thinking as a 'human been' if you don't want to fall like Adam. Jesus Christ also came as a baby and grew into greatness (Luke 2:52).

The Perfect Example

We have to learn from Jesus Christ through how He rejected a shortcut opportunity that was presented by Satan. The devil planned shortcuts to His destiny and tested Him three times after His 40 days of fasting. Without a second thought, Jesus rejected the offers and chose to follow the process. Gamblers love shortcuts because they want sudden greatness that they can't sustain. The long way to greatness is a shortcut.

Many Christians are lazy to the extent that they want God to do everything. All they always look for are miracles. Yes, miracles happen, but God will not do what He has given us the ability to do because He has put success power in the divine laws. Jesus challenged a group of miracles seekers who wanted proof from Him. He told them that no miracles would be given to them.

In other words, they wanted shortcuts in all their lives, but Jesus rebuked the idea.

Greatness takes Time

On your journey to greatness, you need to learn patience, self-development, and persistence. You have got to wait for the grape to ripen before eating it unless you would eat a sour fruit. It is important to consciously develop the right values and attitude when you have nothing so that you can sustain greatness when it is achieved. Whatever you take to leadership reproduces. If you have seen yourself as a CEO or winner of something, the best time to build up the foundation is now. Have you noticed that builders dig down before a skyscraper is built up? Do you also know that every planted seed dies before it germinates? Yes, you have to prepare, create a foundation, and hold the ground before you build. Every great life requires growth.

Before I was inspired to write my first book in 2008, I had almost nothing in my bank account. I was confused like many youths in my country. I thought that was my destiny. On June 18, 2006, the day I made up my mind to take action on my conviction, passion and to live my dreams, everything changed. The point is that things did not change on the outside immediately, but the real me on the inside changed. Then, the next decision was to increase the number of self-development books I read every month from two to three. I began to ask questions, observe, listen and write. I attended many seminars, workshops, and training.

However, it was difficult to discipline myself. I broke my plans several times before I normalized. People laughed at me because I was teaching what no one could see on the outside. Without any hesitation, I kept on doing what I love to do until things started to happen. I was invited to youth meetings to share what I know across Osun state in the Southwest of Nigeria. When my first book came out in 2009, I got more opportunities to speak. Today, I teach peak performance and leadership principles nationally and internationally.

Grow Daily

You may have been working on your goals for years or have been considering what people are saying because you are not yet producing visible results; it is time to focus on your daily growth. People who don't know where you are going in life may laugh at you, but you owe no one an explanation for your conviction. Be in charge of your life and live your dreams. They may neglect you or undervalue you for a while just because you are different. If you compromise day by day, you will lose your dreams.

Many people have given up because of what others are saying about their goals and dreams. If anyone asks you to show your results, tell him or her that, "I am working on something!" Say it like you mean it because you are on your way to the next level. Always remember the story of the bamboo tree.

This natural law cannot be broken by any natural human being. The law is as powerful as the law of gravity. I encourage you to pay today and play tomorrow. You have to decide to succeed and manage it. You manage a decision by doing something about it every day. Intentionally, grow every day. Great people achieve great things daily. Sometimes you need the discipline to move on when you get challenged by situations of life. Give it all it takes. It is not easy, but you have to face your challenges and overcome them. L.I.F.E. is an acronym for; LIVING IN FEAR of EXCELLING. Successful people do what unsuccessful people don't do. You can only become great if you are committed to achieving greatness every day because it is achieved daily, not in a day.

CHAPTER 8:

THE LAW OF LEGITIMATE
SUCCESS

"Real success is achieved as a person, not as an event."

I had a conversation with one of the Liaison Office Managers at the Nigerian Universities Office in Washington, D.C. in the United States, and he talked about how some people secretly get involved in drugs in search of money, which is against the law of the nation. It was shocking and I could as well imagine how people around the world pursue money. People do horrible things to get money because of greed. However, a thousand wrong actions do not make a right. The principles of real success will never change. The fact remains that real success is not what you pursue; it is what you attract. Success is who you are, not what you have. You don't get what you want; you get who you are in life. You don't become great when you are not ready to work out your divine goals and dreams.

Many people make mistakes because they don't know that success as a person is different from success as an event. Those who engage in illegitimate activities to get money only succeed by events, not as persons. As individuals, they are failures; that kind of success never lasts. Only a self-achieved goal and dream is a real success.

Personal Success Requires Lifetime Learning

Universally, one of the reasons why most people don't attract personal success is the lack of continuous learning. You cannot legitimately succeed more than what you know. Have you heard of this idiom before? "What I know is in my brain and what I don't know is in the library". Two top secrets of great inventors and innovators in the world are reading and meditation. They are lifetime learners. Invest your energy, time and resources in great books. If you desire to achieve something worthwhile, you have to read to know and understand. Read to leverage with great minds and know what they know if you want to do something like what they did.

Every scripture was written from the revelation of God. Also, God has revealed the secrets to some people, and they have taken the pain to put them down in great books for you to discover and make your 'long cuts' become 'shortcuts' on your journey to success. They have performed their responsibilities by writing and making them available for you. Your own responsibility is to look for them, sit down, read, understand and apply.

Activate the Winning Habit

I understand that it is difficult to start something new. In fact, anytime we want to have changes, our minds go into turmoil. Some habits are easily formed and hard to be broken because it may be difficult to develop a habit, but once developed, breaking it is hard. Ignorance becomes your number one enemy when you imbibe a learning habit. If you are not used to books just like I was, begin by reading at least a page every day and discipline yourself to do it consistently for 21 to 30 days. You only need decisions and management of these decisions to follow through with your resolutions. Then, your next stage would be to read at least one book in a week. It is possible to grow speedily and start to devour great books in your chosen career area.

As you choose to develop every day, welcome to the world of great minds as you learn something new daily. Sam Levenson, who was an American humorist and journalist said, "Don't watch the clock; do what it does. Keep going." Once you start, keep growing and always remember that readers are lifetime learners, and learners are potential leaders. Read, discover, think, innovate, invent, and lead. Your real success will soon be discovered and celebrated.

CHAPTER 9:

LAW OF ACCURATE
COMPENSATION

"You equally get in life, whatever you give to life."

This law is all about cause and effect. Every great thing in the world is caused by someone or a situation. Many people want to experience effects they haven't caused. You should not be receiving what you are not giving. The foremost reason why you have harvested nothing is that you have sown nothing. It is a law.

Giving Precedes Receiving

We like to receive things that we cannot give. Sometimes, we want love from people but we don't demonstrate love. We want money without having nothing to exchange with it - no products, no services. If you dare to extend the hand of love to people, somehow, the natural phenomenon designed by God will arrange great measures and bless you accordingly (Luke 6:38). Give and take is the language that life understands. The quality of input is also the quality of output. Jesus said, "I have the power to lay it down, and I have power to take it again" (John 10:18). We all have the power to give and receive but don't expect to take it again if you have not laid it down.

Let me give you a practical experiment; if you shout in a forest whilst hiking up a mountain, the same voice, I mean your voice will shout back at you. That is an echo. It is also called life. What you send to life is what life sends back to you. It is a law of nature. Now, do you speak and hear negative words, feel and do negative things and expect positive feedbacks? Are you so selfish that you think of yourself alone? Do you frown at people, but expect them to smile at you? It doesn't work that way. The moment you disconnect from the flow of give-and-take; you will undoubtedly experience a stopover of blessings until a connection is made again.

Solve Problems for People

When you identify opportunities that present themselves as challenges, don't run away from them because they are meant for you. They are gold-wrapped for you to unveil. It pays you to decide today on the problems you can solve for people, which they can pay you for. Your purpose and passion are connected to problems awaiting solutions in your hands. What is your talent or natural skill that is crying for expression? What makes you sing, cry and dream? Think of solutions to what people call problems in your family, community, city or state. The same problems are common to almost everyone in the country, continent and the world over. You may likely be the one to solve a national problem if you start in your family with what you can do. Try to make a change where you are. You cannot be committed to solving people's problems if you don't have the heart of giving. When you dispense your value, you will achieve real success. Nature finds a way to compensate you accurately when you solve a great problem. Remember that God created this indispensable law and nothing of Him is unreal.

Life Pays Accurate Compensation

Zig Ziglar, who was a renowned American author and motivational speaker said, "You can have everything in life you want if you will just help other people get what they want." The secret is to give people whatever you expect life to give back to you. Steven Covey describes 'Emotional Bank Account' in 'The 7 Habits of Highly Effective People' as a withdrawal and deposit of emotions. When you love or hate people, you deposit love or hatred. When you deposit love or hatred, then you will withdraw the same. You don't withdraw from a bank without having deposited money in your account. You only request for a withdrawal when you have deposited. If you present a perfectly filled cheque to life without any deposit in your bank account, it will bounce back to you. You cannot harvest what you have not sown. In other words, life doesn't give you what you haven't released because you don't deserve it. I encourage you to help, forgive, celebrate love, and pray for people. You only get what you give.

CHAPTER 10:

THE LAW OF COMPASS

"If you can create a roadmap by organizing your thinking, you will produce an outcome."

Count the Cost

It is not enough to discover your purpose and dream alone, it is also important to plan to achieve them. You can have a big dream and still be frustrated and depressed if you don't organize your thoughts strategically. The scripture teaches, "Suppose one of you wants to build a tower. Won't you first sit down and estimate the cost to see if you have enough money to complete it?" (Luke 14:28). You need to know when and where to go each time. You have to understand analyses like who, when, what, and how in every situation. Planning gives you direction and enhances your focus. You need a to-do list that must be followed daily.

Organize Your Thinking

I could remember my days as a Corps member during the compulsory one year National Youth Service Corps in Nigeria; I decisively applied the principle to direct my focus. It was difficult for anyone to interfere with my organized thoughts or involve me in activities that were far from my plans for the year. I set some short term goals that were clearly written in that year's diary and designed strategies despite the challenges. There were activities from the program that I couldn't control, but I was able to discipline myself by investing five minutes every morning to organize my thoughts. The plans I sketched after two months of the service year positioned me to achieve a lot within the shortest time and I eventually got accolades including a national medal.

Has anyone built a skyscraper without having a plan? The answer is no. The higher the building, the longer the time it takes to plan the foundation on paper before the work begins. Planning is important in the process of constructing a long-lasting and great building. The same way the law of planning is compulsory for you to obey if you want to achieve great things. You need to organize your thinking intentionally. When you plan, you prepare for the known and unknown.

Strategies

How bad enough do you want to succeed in business and in life? If you want a change in your life, think and plan. People fail when they fail to think and plan. Successful people think and plan to achieve their dreams. Planning gives your life or business a direction to be focused on from the foundation.

Once you have known why you are living, organize your thinking to achieve it. God will not descend from Heaven to plan for you. Therefore, if you want to live your dream, deliberately plan to position yourself in the right place where you can develop in that direction every day. The moment you start to work on your dream, you can begin to live intentionally. But if you do nothing about your dream, then you will continue to make a living. It is a choice. All you need is to be on purpose because you have all it takes to explore.

You have to think and plan daily. The moment you fail to plan, you have invited frustration without knowing.

Dare to plan and don't give away your life for someone to control. You may start small but plan the big things. You may begin your journey by working for someone, but plan to employ others. You are special and you can also be in charge because you have greatness in you. The future is open for you to manifest. The identified disaster is that if you don't plan, you will always follow another person's plan for the rest of your life. If you don't plan your life, other people will do and the benefits will not come to you.

Think about Thinking

Sit down and think about thinking. Take a walk, ask questions, listen, observe and meet people. Release your mind and plan big. Take a pen, write down the ideas and plan to act without procrastination. You have all you need to fulfill your destiny inside you.

I encourage you today to sit down and plan. Why? Your dream may be real, but what gives it breath is your strategy. Planning is the personal management responsibility between your conception and destination. Aim at something so that you can hit it. Don't forget that if you also aim at nothing, you will hit nothing. Let your plan be big. It is better to plan big for the future and not achieve all than to plan small and achieve all. Invent into your purpose with great plans and see God exceed it.

CHAPTER 11:

THE LAW OF POSSIBILITY

*"When you disbelieve the impossible, you will create the
self-belief for the possible."*

Nothing is Impossible

Most times we behave like Thomas in the Bible. Despite his journey with Jesus Christ, he still had an unbelief approach to Jesus' resurrection.

He did not believe that Jesus rose and showed Himself to other disciples until He appeared to him and saw Him physically (John 20:29). The only way to make God helpless in your life is when you have no belief. Belief precedes miracles (Matthew 21: 21-22).

You can succeed in anything because nothing is impossible. Everything, including what you are thinking right now is possible if you believe. Impossibility does not exist in God's agenda for individuals, families, organizations, or nations. You can break free from a paycheck job and impact the national economy with your business idea. You can be the next inventor or innovator in your field. You can break free from poverty and become wealthy. You can break free from the devil and become God's friend. You can be totally released from the prison of the mind and be transformed. You can create products and services that can move all over the world because you believe you can. You can make things happen and become what you have been designed to be on Earth. Henry Ford, the founder of the Ford Motor Company said, "If you think you can or think you cannot, you are right."

Don't Believe Them

Never allow anyone to tell you that something is impossible. There is nothing named 'can't'. People only complain about success and criticize it because they can't achieve it. Don't believe the word 'can't.' Let me share a story of a heroic British man, Sir Roger Bannister. It was universally believed that it was impossible to run a four-minute mile until 1954. But Roger mentally disengaged himself from that belief. He showed the world a new perspective of possibility when he broke the four-minute barrier by running it in three minutes fifty-nine and four-tenths seconds. You too can make a mark that cannot be erased, if you believe.

We all need to develop our inherent belief system. This is what we conceive about ourselves when we are alone. It is good and easy to tell people about your business and personal life goals. It is wonderful to declare that you are a world champion with hope in the midst of friends. People tend to see you like a different being. Though, nobody celebrates you until a result is seen. The greater part of every great reality is done while alone. Silently speak to yourself. Strong possibility thoughts that drive actions are developed when you soar alone on the inside like an eagle.

Neutralize Negative Thoughts

Take a look at yourself in the mirror. The mirror enables you to stand alone and speak to yourself that you can achieve before you believe. Ignore your zero or average background and tell yourself that "I can make it," "It is possible to achieve my business goals and dreams," "I am a world champion," "I disbelieve, I can't," "My life can change," "I will be great," "I can invent and innovate," "I am a leader," "I can touch my generation," and "I can fulfill my destiny". Intentionally neutralize your conditioned inner wrong belief by speaking and hearing positive words. Intrinsic belief precedes uncommon actions. The Bible indicates, "For as he thinks in his heart, so is he" (Proverbs 23:7). Basically, you will become what you think. Always speak to yourself.

I encourage you not to allow someone's negative words to position you for sickness and weakness. When you hear words of discouragement about your dreams, choose not to make them yours. Your decision at such moments is very important to the achievement of your dreams. You have the seed of greatness in you and you can achieve your dreams. If people don't see things working for you right now, very soon they will be amazed because your dreams will come to pass. Just keep on doing something about your goals daily. Always remember that opinions are like households and everyone is entitled to one. Don't accept someone's opinion of the impossibility to become your reality.

CHAPTER 12:

THE LAW OF CHALLENGE

"Every promotion requires a successful examination."

Face the Music

Why do we really hate what we call problems? Who else do you want to have your problems if not you? We don't want challenges, but we want to experience promotions and achieve greatness. No solution comes without a problem and no problem arises without a solution. Challenges precede promotions. As human beings, we naturally don't want to be troubled, but challenges are inevitable. They are part of life. We cannot run out of life. It is time to face the music.

Challenges are moments of self-renewal and joy (James 1:2). An eagle understands this better. When an eagle is about 40 years, its feathers become weak; it flies to a high mountain to renew its feathers. There, it removes its feathers of about 7,000, depending on its size. The eagle hits its beak on the rock; there it stays for 40 days to renew its strength. Sincerely, the process is always painful for the eagle but it has to do it and live for more years. This gives an eagle the longest life-span of its species. Great achievements are attached to solving great challenges. The process can be daunting, but it's worth growing through.

Go through the Process

Behind every breakthrough lies a challenge to conquer. Great people on Earth passed through great examinations one time in their lives which they overcame successfully. Think of Nelson Mandela who was a South African political prisoner for 27 years with hard labor in Robben Island prison before he was elected as President of South Africa in 1994. Think of Thomas Edison who failed 9,999 times, before he was able to complete the invention of the incandescent bulb; think of Joseph in the Bible who experienced terrible challenges and eventually became a Prime Minister at the age of 30. Think of David who killed Goliath before he was later exalted to be a king of Israel at the age of 30. Think of Jesus Christ who was killed without any sin before He was glorified and seated at the right hand of God. There are many other great people who had challenges and succeeded in the Bible and in the world. Attacking our challenges and winning is worth it. They are there for us to grow through and conquer, not to go through and get depressed.

The greatest challenge I successfully won in my life was to believe in myself. It was a battle because everything around me did not present hope until I could understand the power of winning, no matter the circumstances. Your life challenges right now are there for you to grow through and become great. They are like fumes that will soon fade away. Don't run away from reality because every promotion requires a successful examination. Every challenge you face on your journey of success is an examination you need to pass. Sit down and write the examination. Face the music. If you don't have an answer, search for it. Ask questions, observe, and listen. The solution exists.

Be a Problem Solver

Your answer to a problem precedes your breakthrough. Think about your goals and dreams, problems are seated at every transition for you to solve. People only recognize you for the problem you have solved. Leadership positions are principally meant for problem solvers. The level of the solution you are capable of providing depends on the amount of money you receive in exchange.

The first thing Jesus did in His village was to solve Peter's problem. He solved his problem of not catching any fish for several hours despite Peter's expertise, and Peter followed him. Jesus was looking around for problems to solve. People don't follow you until you help them to succeed. The first thing God does when you pray for a great thing is problem shopping. God will shop for an equivalent problem to your prayer requests for you to solve.

What you focus on will always determine what you see. Always see the positive side of every challenge, and move on. Sometimes it is hard to speak for yourself, but you need to do it.

CHAPTER 13:

THE LAW OF ASSOCIATION

"Every relationship increases or decreases you."

Run with Champions

This is one of the unbreakable laws that many of us are not conscious of. Obedience to this law can shoot you to success because success is attracted when this law is obeyed.

Where you plant and grow your seed of greatness matters. If you plant it among thorns, it would shrink and either produce small fruits or die. People are experts in what they have passed through in life. Those who are hurt are excellent in hurting people around them. They only teach what they know best because you can only give what you have. Those who choose not to give up and then succeed are brilliant at encouraging people to persist. If you run around with losers, you will end up a loser (Proverbs 13:20). Hans F. Hansen, a retired international football player said, "People inspire you, or they drain you – pick them wisely."

Who is Eating You?

Some people are pathogens while some are pests. Pathogens consume plants from the inside out, but pests consume plants from outside in. People who are both pathogens and pests are enemies of greatness. You can physically see some like pests as they eat you, but some are invisible or far from you but still eat you all the same. Now, the question is, "Who is eating you?" If you are always around four negative people, you will end up being the fifth one. Your belief system is always programmed through what you hear and see consistently. It is highly important to choose how your eyes and ears work. There are some people who are only creative in inventing reasons why things won't work. Unfortunately, some are your family members and close friends. They would itemize people who have attempted the same thing and failed. They complain and moan all day. They are dangerous to the fulfillment of your dream. Fortunately, you have the choice to listen and believe them or not. What you believe is what you will become.

Valuing People

This law is about valuing people and intentionally choosing people you walk with. The moment you discover where you are going in life, connect with like minds. Greatness is impracticable in isolation. During my master's degree program, I encountered a lady who hardly spoke with anyone. She ignored people who greeted her. In fact, the first time I met her, I felt embarrassed because she looked at me as if I was nobody, even though I didn't look shabby on that day. The lesson is this; if you work or live in a place without friends, you are a social orphan. You need people to achieve your dream. You cannot achieve your dream in seclusion.

You are where you are right now for a reason. You have to be conscious about connecting with the right people. You do not know the person God wants you to meet now. Some people would meet God in person and never welcome Him. The Creator performs miracles in people's lives through human beings because man holds the Earth's license. God has given us the power to dominate the Earth. God designed the world for all men and women to be in charge. We were not designed to live outside the Designer's concepts. The Earth is our territory and we need to understand the power of relationships in it. Always recognize people because every human being is an extension of God.

Move with only Quality People

Let me tell you a story I once read about the eagle who thought he was a chicken. There was an Indian brave man who walked down a beautiful trail where he discovered an eagle's egg that had fallen out of its nest. He picked the eagle egg but realized as he looked up that the nest was too high for him to return the egg. He decided to put the egg in a nearby prairie-chicken nest where the egg also hatched. The young and beautiful eagle grew up with the other chickens. In fact, what the chicken did, the eagle also did. The eagle kept moving around with the chickens because he thought he was like them. The chickens flew short distances and the eagle also did the same, since the chickens could not fly long distances. The eagle learned and reproduced the chickens' habits. He thought that was what he was supposed to do. The eagle thought that was all he could do, and that was all he was able to do.

One great day, the eagle saw a bird flying high above him. He was wholeheartedly encouraged. Then, he asked the hens, "Who is that?" The hens said, "That's the eagle, the king of the birds. He belongs to the sky, we belong to the earth – we are the chickens." Ignorantly, the eagle lived and died a chicken because he thought he was. What a great story. Can you imagine?

This is how many people lived and died because of wrong relationships. You don't deserve to die like a chicken. If you want to achieve your dream, you have to leave the chickens' yard. Chickens don't soar; they groan and complain in their yards. Eagles soar. You need to live a full life, fulfill your destiny, and die empty. If you could check your relationships and ask yourself the following questions, you will help yourself a long way. "Of what benefit exactly is this relationship doing to me? Am I growing or dying every day because of this relationship?" Once you provide answers and then take the necessary actions, things will change.

Sometimes, in your business, some employees are not worth it. Stop managing them. You have to work with great minds if you want to grow a great company. The relationship is the key to development. Your business and your life cannot be better than the quality of people around you.

Become a Winner

You have to move with winners if you want to become one. Mingle with great minds and not small minds. Small minds cause great problems, but great minds cause small problems. Great people are committed and addicted to great relationships. You cannot dine and wine with negative people and become a champion. There is no neutral ground in every relationship; you are either added to or removed from. Every relationship increases or decreases you.

CHAPTER 14:

THE LAW OF FEAT

"Every extraordinary achievement calls for an extra effort."

Greatness is a sweet experience, but it requires extra efforts. It necessitates self-responsibility and tireless pursuit because it cannot be inherited. Those who want more always do more. If you want to achieve more, you need to do more. Do you really want to become a real leader and achieve greatness? Then, live your dreams everywhere you go by doing these two things daily: build a difference and make a difference. You have to want it bad enough.

Do More

In a fruit juice producing company, Felix, a new senior officer, who was employed to purchase ingredients for the company was sent with the driver to make a purchase for the next production. After he had worked for two years in the organization, he started praying for a promotion because he was denied one when it was to be given. He did not know why. One day, the department head discovered through the newspapers that the prices of fruits dropped. He decided to send Susan, a junior staff with a different driver for the same purpose. Having gotten the prices the next day, Felix returned with a list of six fruits as he usually did. Susan also returned with a list of different ten fruits and separated prices as per one and in bulk. She used her money to buy a sample of the fruits each and also took pictures, and this was used to make her report.

When Felix and Susan made their presentations, it was obvious to all that Susan was more creative and innovative. They were both sent on the same assignment, but Susan did more than what she was sent to do. After four years in the company, her attitude helped her to get two promotions. However, Felix did not receive any promotion and was later sacked because his attitude to work couldn't move the organization forward. This is the point: always be creative and innovative to do more. Organizations employ and pay people to solve problems, not to create them. Do more than what you are sent to do in every situation. If you want to achieve extraordinary things as Susan did, always think through on how to add value and get more results.

Destiny Demands an Extra

Destiny demands an *extra*. I would suggest that you write this on a plain paper and paste it to where you can see it every day. To achieve your vision, the process always calls for more action. The *extra* positions a man or a woman to look like someone that is different from others. Doing more attracts great opportunities and positions of leadership. You cannot be someone who always goes the extra mile and not influence people positively.

Every true vision demands for energy. How badly do you really want your vision? You cannot desire greatness and not have extra energy. Leaders are energetic people. They do more when others are tired. Jesus is a perfect example. His spiritual power, energetic prayer, and extended mind distinguished Him from other human beings on Earth. He did more in the power of God. Jesus valued people that others couldn't love.

His love was impeccable and extraordinary. Love was His tool in achieving His vision on Earth as given by the Father. His attitude toward people attracted others to Him. If you also love more as you live your dream, you will become an extraordinary person.

The Leadership Call

Leadership positions are not meant for people who easily give up. They are for people who are ready to take responsibilities, persist, and do more as they work on their visions. No stopping anyone that has the habit to persevere. He or she grows into leadership.

The leadership call requires the right and unusual actions of ordinary people, and these produce rare results. Leaders don't just solve problems; they accommodate and solve more problems. They have the understanding that pressure refines; gets rid of what God has not put in their lives and produces gold. Lives that display gold are always at the leading edge.

If you want to start leading from where you are and experience greatness in life, I encourage you to intentionally keep doing more than whatever you are asked to do. Put in the *extra* into what you do and see yourself achieving extraordinary results.

CHAPTER 15:

THE LAW OF FOCUS

"The most powerful indicator in getting the desired result in every situation is to neither look to the left nor the right on the way forward."

The summary of this law is: 'I attract to myself whatever I give my focus, attention, or energy to; whether wanted or unwanted.' This principle applies to our lives and everything on Earth. It is impartial and impersonal, which means it works when you want it to and when you don't want it to. You invariably achieve the best from what you focus on, either good or bad. The question now is, "What are you focusing on right now? Your vision or what people want you to be?" "Are you improving lives with what you do daily?" Your focus must be defined and fixed.

Stay Engrossed with Passion

If you focus on your immediate environment and could recognize yourself as being broke, poor, or lonely, you will eventually believe the thoughts, and guess what? That is exactly what you would be. If you believe and focus on "I can", yes you can. If you also believe and focus on "I cannot", yes you cannot. You need to change your focus if all you do now is not taking you towards Canaan land.

In a foul Roman prison, Apostle Paul didn't lose his sense of passion, direction, and command of his environment. Despite his horrible condition, he preached focus, priorities, goal-setting, and perseverance. He made it known that it is imperative for us all to let go of those things that don't matter (Philippians 3:5-9) and focus on our primary assignments in life. To move forward and keep moving forward, you have to forget what lies behind and look forward to what lies ahead. Pressing forward is a non-negotiable habit for anyone who wants to fulfill a purpose and achieve greatness.

Focus helps to Recognize and Maximize Opportunities

When we focus on every situation, neither looking to the left nor the right on the way forward, we usually recognize and maximize opportunities. Opportunities abound, but only a few people recognize them because of 21st-century sources of distraction. Everyone wants to move to the next level, but most hearts are distracted and closed. I was once on a plane flying from London to Ethiopia when I engaged some British businessmen in a discussion on their knowledge about Africa. One of them said that Africa is a fast-growing continent and they have kept their expertise focused on available opportunities. Focus helps us to recognize and maximize opportunities in our immediate environments. Some Africans travel to western nations to survive, not to explore because they have lost focus of their vision. Do you have your inner eyes open to recognize problems that you have the power to solve with your gift? How much information do you have about people and situations around you? Answers to the two questions are crucial toward fulfilling your dream.

Focus Triggers the Next Level

Lack of focus makes us misplace priorities, disengage from responsibilities, and stay longer in a condition. I would like to encourage you to take up responsibilities when you recognize them. David possessed this attribute and got remarkable results. He focused on the use of his gift and that activated his next level. The 'Sling' was his gift that he used to conquer the lion and the bear. It was also used to kill Goliath. He never lost his focus on using his divine gift to produce supernatural marks. The same opportunity everyone ran away from was the one he embraced. When he had the opportunity, he responded and took ownership. This attitude led him to a greater level. You too can move to the next level if you focus on the use of your gift and recognize opportunities when they show up.

So, we need to concentrate on our dreams and open our spiritual eyes to recognize people's needs. You need to be sensitive to problems that are opportunities around you. You need to see what others can't see and make marks that others won't make as we take charge. If you stay connected to your vision, you will become what God wants you to be.

CHAPTER 16:

THE LAW OF MENTORSHIP

"Your journey to success is quickened when you stand on the shoulder of an achiever."

I have learned that we all have God's given potential to create what we want in our environments. How you think and behave is dependent on your environment, and your future is dependent on how you think and the actions you take. We have the power to choose people who have the capacity to create the environments that suit our future. Why? Because you become like whom you invest your time with.

Submission under mentorship is the understanding of the 'wise learning strategy' (Proverbs 1:5). Quickly note that this is not a 21st-century strategy. It is one of the spiritual processes that God prepared for us to be discovered, understood, and obeyed on our journey to greatness. What I call the 'wise learning strategy' is serving and receiving thorough guidance under someone who is a leader in your area of divine calling. Being connected to a source of good nutrients is highly important to achieving greatness.

Connect and Achieve More (CAM)

The acronym 'CAM' crossed my mind for the first time in 2012 while speaking about mentorship to Corps members during their three weeks camping program in Nassarawa state, North Central Nigeria. I have been inspired by different accounts in the Bible on mentorship. Think of Moses, who mentored Joshua; Apostle Paul, who mentored Timothy; Elijah, who mentored Elisha and so on. In fact, Elisha received a double portion and God doubled the miracles he performed through him.

Let's look at the fantastic relationship between Elijah and Elisha. *"And it came to pass, when the Lord* was about to take up Elijah into heaven by a whirlwind, that Elijah went with Elisha from Gilgal. Then Elijah said to Elisha, "Stay here, please, for the Lord has sent me on to Bethel." But Elisha said, "As the Lord lives, and as your soul lives, I will not leave you!"

So they went down to Bethel. The sons of the prophets who were at Bethel came out to Elisha, and said to him, "Do you know that the Lord will take away your master from over you today?" And he said, "Yes, I know; keep silent!" Then Elijah said to him, "Elisha, stay here, please, for the Lord has sent me on to Jericho." But he said, "As the Lord lives, and as your soul lives, I will not leave you!" So they came to Jericho" (2 Kings 2:1-4). Here, Elisha took a big risk. He made a lifetime decision to follow Elijah. He kept watching great Elijah for years as a protégé. He was so close to Elijah that he had the privilege to ask and receive the double portion. Elisha had a servant's heart and waited for God's time of anointing to replace Elijah and do more. Every emerging leader needs mentors. If you want to lead and be great, you must be hungry to grow and pursue great mentors.

People's Impact on Vision

Mentorship is fundamental if you have a real vision. You have to find someone who has been what you want to become. Nobody has the capacity to achieve a big vision alone without the support of someone or people. Nehemiah, who received a big vision to rebuild the wall of Jerusalem, had to speak to the Jews, priests, nobles, officials, and others that, *"You see the trouble we are in: Jerusalem lies in ruins, and its gates have been burned with fire. Come, lets us rebuild the wall of Jerusalem, and we will no longer be a disgrace"* (Nehemiah 2:17, with emphasis). Nehemiah had to call other people to help him achieve his God-given vision. God has prepared someone or some people to work with you and help you to achieve your big vision. You need to see with your inner eyes, observe, study, listen, and ask questions if you want to find them.

Wine and Dine at the Feet of Greatness

The relationship with mentors can be challenging and sweet. Such relationships make life easier. Wow! Is it possible to navigate life and avoid certain pitfalls? Yes! A mentor makes your life easy because he or she helps you make a long journey a shorter one. He or she stands as a ladder that can speed up your journey to the top. You can achieve your vision faster than you can ever imagine. Actually, you have to obey this principle if you want to achieve greatness. You cannot become mediocre if you dine and wine with great people.

You have to drop personal pride if you want to follow a mentor and submit to instructions. Then, you have to do this willingly. You cannot be pushed up a ladder unless you are willing to make moves step by step from the bottom to the top. Once you have decided to move up, position yourself very well in order to enable mutual expectation, mutual contribution, and mutual commitment. Both Elijah and Elisha expected to achieve great things for God. In fact, Elisha had an expectation of receiving a double portion of anointing from Elijah.

Sometimes, your mentor could be someone that you have never met before but whose achievements you desire or someone you are familiar with. If the person is far from you, get his or her books and CDs. But, if you know the person physically, connect, and for you to achieve this, never enter the relationship empty-handed. Go into a mentor-mentee relationship with value. The issue here is that you must connect, and that may not necessarily be through a physical relationship. If you know someone whose results you desire, first endeavor to clarify the purpose of the gains that you want. Focus on learning and practicing. You cannot be committed to this and not succeed.

Differentiate Sources of Knowledge

No matter how much you think you know, you are yet to know everything you should because someone has more information. Differentiated knowledge clarifies that learning could be varied. Mentorship doesn't go with age; it goes with knowledge, experience, and power. It could be 'up', 'lateral', or 'down' mentorship. A mentor may not be someone who is older than you. It could be someone of the same age or younger than you. Don't allow pride to disconnect you from your learning databank.

It was difficult for me to learn from people of my age bracket and the younger ones until I discovered that fighting self-pride and embracing humility to know what I didn't know is imperative in becoming a real leader. When I discovered this, I decided to choose friends who are going in the same direction as me. Now, who are the people that you spend time with? Ask yourself, "What are these people doing to me?" "Are they adding to me or subtracting from me?" "What do they want me to settle for?" The last question is the most important one because people who are close to you can help you to settle for the average. To achieve greatness, keep company with those who are going to help you on your journey, not those who are going nowhere.

CHAPTER 17:

THE LAW OF ACTION

"Things move when you cause them to move."

Don't be a Wisher!

Mr. Myles Munroe said, "Planning without action is futile and action without planning is fatal." I want to quickly remind you that you need to take action on your set goals. A dream without action remains a wish. The difference between a dreamer and a wisher is 'action'! Dreamers act because they believe the dreams wholeheartedly and they know that these things will come to pass. If you have great dreams, an unshakable belief that all things are possible, and with assurance, but without action, you will experience a result called 'nothing' (James 2:14-17).

The only proof to show to the world that you believe in your dreams is in your actions. If you have an idea, remember that you may not be the only one thinking about it in the world. After the generation of an idea, analyze it and execute it. Your idea does not rule the world until it blesses the world. If you have an idea, act on it. There is no difference between someone who does not know and the one who knows everything without corresponding actions. Walt Disney said, "The way to get started is to quit talking and start doing."

Just Do It

In 1960, John F Kennedy had a challenge as the President of the United States. The Russians were the first to put a man into space. On a great day, he invited everyone that mattered to the White House in order to debate on how to get to the moon before the Russians.

After several deliberations, no idea could help. Then, a brilliant German Scientist, Neil Armstrong suddenly told them that he knew what to do. Neil Armstrong spoke out the five words that took the nation to the moon. He said that all we need to make this happen is 'The will to do it'! This statement provoked actions. These five words commanded silence in the room and prompted the President to sign billions of dollars for the historic journey. It became a 20th-century history that in 1969, as part of the Apollo 2 mission, Neil Armstrong was the first man to achieve the dream of walking on the moon. All you need to live your goals and dreams is 'The will to do it'.

I am very sure that you have something in you. You have something different from the next 10,000 people around you. You have something that the world has never experienced before. You have the capacity to touch the world. You have the seed of greatness within you and you can make things happen. You have the ability to turn the thousands of ideas that are flowing in your mind daily into innovations and inventions. Your action confirms your attitude. When you act, you break traditions and produce results. Just do it!

BONUS:

USE OR LOSE IT

I enjoyed riding with my father in his first car when I was growing up. One sunny afternoon, I had a rethink and decided to learn driving when he traveled because he parked the car for three days and I couldn't drive it. When he returned, he wanted to take the car to an urgent church meeting the same day but the ignition refused to respond. He quickly invited his mechanical engineer, and the Engineer revealed that the ignition did not work because it had not been used in three days. He was disappointed and decided to avoid the meeting. This same experience happens to every human being in this world. The principle is *'use it or lose it.'* No matter how gifted you might be, if you don't use it, you will lose it. Those five words can create frustration, anxiety, poverty, wealth, fulfillment, joy, sadness, spiritual development, greatness, leadership opportunities, and many more. Your result depends on your attitude toward your gift.

Servicing your gift every day is an unavoidable process if you want to lead in your domain of gifting. If you neglect your gift without daily servicing, it will disappoint you when you need it the most. This experience is common around us, but the fact is that what you do with your seed of greatness determines your own experience. God who put the seed of greatness in you finished the work by accompanying it with the grace of fulfillment when used. If you don't want to lose your value, don't cease functioning.

Every effect is caused. If you want to become someone to be followed, you have to cause it to happen. Leaders service their talents daily. One thing I know about you is that you have books, movies, songs, businesses, and inventions in you. You are the only one that can bring them to life and put your signature on them. Breathe into what you have in you and give it life. Search for your gift, your talent, and your passion, and work things out of them. God has said it and I am convinced about it that your gift will make a way for you in the world.

Now that you are equipped to start living a life of exclamations, not explanations, you've got the power to become the leader that you are meant to be. Then, if you can obey these 17 indisputable divine laws, get hungry with your dreams and maintain your anger, persist during tough times, do what ordinary people won't do, practice servant-leadership by serving your gift to people, pray and trust God for breakthroughs, believe me - the wall of possibility around your future is built and greatness will become inevitable for you.

Keep on leading!

RESCUE TEAM PROJECT (RTP)

... Everyone is a leader!

Vision: Raising leaders in African Nations.

Our Mission Mandate: To help Africans discover, develop and demonstrate their true potentials.

Our Dedication: Rescue Team is dedicated to transforming followers into Leaders by equipping individuals with our leadership beliefs.

Rescue Team Project (RTP) is an African nations' leadership development mission. It was inspired through a passion and ignited by a purpose to help people discover, develop and demonstrate their leadership potentials. We are committed to damaging mediocrity. RTP believes that '*everyone is a leader.*' Every African nation needs leaders, not position occupants. Position occupants focus on power and the next election, while leaders focus on empowerment and next generation.

Project Motivation

Our history in Africa is based on oppression through colonization, which has formed our daily way of life. The operation was so intensified with the capacity of producing different destructive national philosophies, which are enormously difficult for us to break. These beliefs have trained us not to be able to express ourselves, vent our thoughts, and live our originality.

However, leadership is not about someone else, it is about everyone. Leadership is everyone's business. Every human possesses leadership potential because everyone is gifted. Everyone has the leadership seed on the inside. If we don't embrace these beliefs now, our children and the next generation will be victims of other people's initiatives. It is high time we renew our minds, change our thinking and get the understanding that we all have leadership abilities to influence everything we can think of.

Our Beliefs

- Everyone has leadership ability (Everyone is gifted).

- Leadership is born when a gift (leadership seed) is discovered.

- The leadership seed must be developed (Leadership is a journey, not a day trip).

- Create a product or service (value) from your gift.

- The capacity of your product or service distribution determines your leadership level.

- You lead when you dominate your domain (work territory), not people.

RTN was officially launched in Nigeria on December 20, 2014. Nigerians need to understand that we all have the power to influence anything. Our actions or inactions influence people daily. We have different potentials and these are resources for national development. We have the capacity to innovate and invent great things and export to other nations. And because everything rises and falls on leadership, the government is meant to be the vehicle for national development. In Nigeria, this project is focused on producing real leaders in families, communities, local governments and states.

To make inquiries or support this vision, visit: www.waleadekanla.org or send an email to everyoneisaleader@rescueteam.com.

ENDNOTES

Each indisputable law was created by the author from texts and contexts in the Bible and detailed below:

CHAPTER 1

1. THE LAW OF GREEN THINKING: *"You naturally grow living thoughts when you open your mind to worthwhile things."*

   Source: Proverbs 23:7

CHAPTER 2

2. THE LAW OF DISCOVERY: *"You don't create the purpose for your creation, you discover it."*

   Source: Proverbs 19:21

CHAPTER 3

3. THE LAW OF SERVANTHOOD: *"If you hold service as a key, you will open the door of greatness."*

   Source: Matthew 20:20-28

CHAPTER 4

4. THE LAW OF PASSION: *"You get the best out of life when you are positioned in your zone of strong thirst."*

   Source: Matthew 6:21

CHAPTER 5

5. THE LAW OF INSIGHT: *"Whatever your inner eyes can see is what you can get."*

    Source: Jeremiah 1:11 and 13

    CHAPTER 6

6. THE LAW OF PROPHESY: *"You command new things into existence each time you speak."*

    Source: Genesis 1:3-29

    CHAPTER 7

7. THE LAW OF STEADY GROWTH: *"Daily success accumulates to become an endorsed success."*

    Source: Luke 2:52

    CHAPTER 8

8. THE LAW OF LEGITIMATE SUCCESS: *"Real success is achieved as a person, not as an event."*

    Source: Joshua 1:8

    CHAPTER 9

9. LAW OF ACCURATE COMPENSATION: *"You equally get in life, whatever you give to life."*

    (Source: Galatians 6:7)

    CHAPTER 10

10. THE LAW OF COMPASS: *"If you can create a roadmap by organizing your thinking, you will produce a ground-breaking outcome."*

    Source: Luke 14:28

CHAPTER 11

11. THE LAW OF FOCUS: *"The most powerful indicator in getting the desired result in every situation is to neither look to the left nor the right on the way forward."*

Source: Philippians 3:14 and Proverbs 4:25

CHAPTER 12

12. THE LAW OF POSSIBILITY: *"When you disbelieve the impossible, you will create the self-belief for the possible."*

Source: John 20:29

CHAPTER 13

13. THE LAW OF CHALLENGE: *"Every promotion requires a successful examination."*

Source: James 1:2

CHAPTER 14

14. THE LAW OF FEAT: *"Every extraordinary achievement calls for an extra effort."*

Source: Philippians 4:13 and Colossians 3:23

CHAPTER 15

15. THE LAW OF MENTORSHIP: *"Your journey to success is quickened when you stand on the shoulder of an achiever."*

Source: Proverbs 1:5

CHAPTER 16

16. THE LAW OF ASSOCIATION: *"Every relationship increases or decreases you."*

Source: *Proverbs 13:20*

CHAPTER 17

17. THE LAW OF ACTION: *"Things move when you cause them to move."*

Source: *James 2:14-17*